W9-CLR-558

URANUS

U R A
The Seventh Planet
A Voyage into Space Book • Franklyn M. Branley

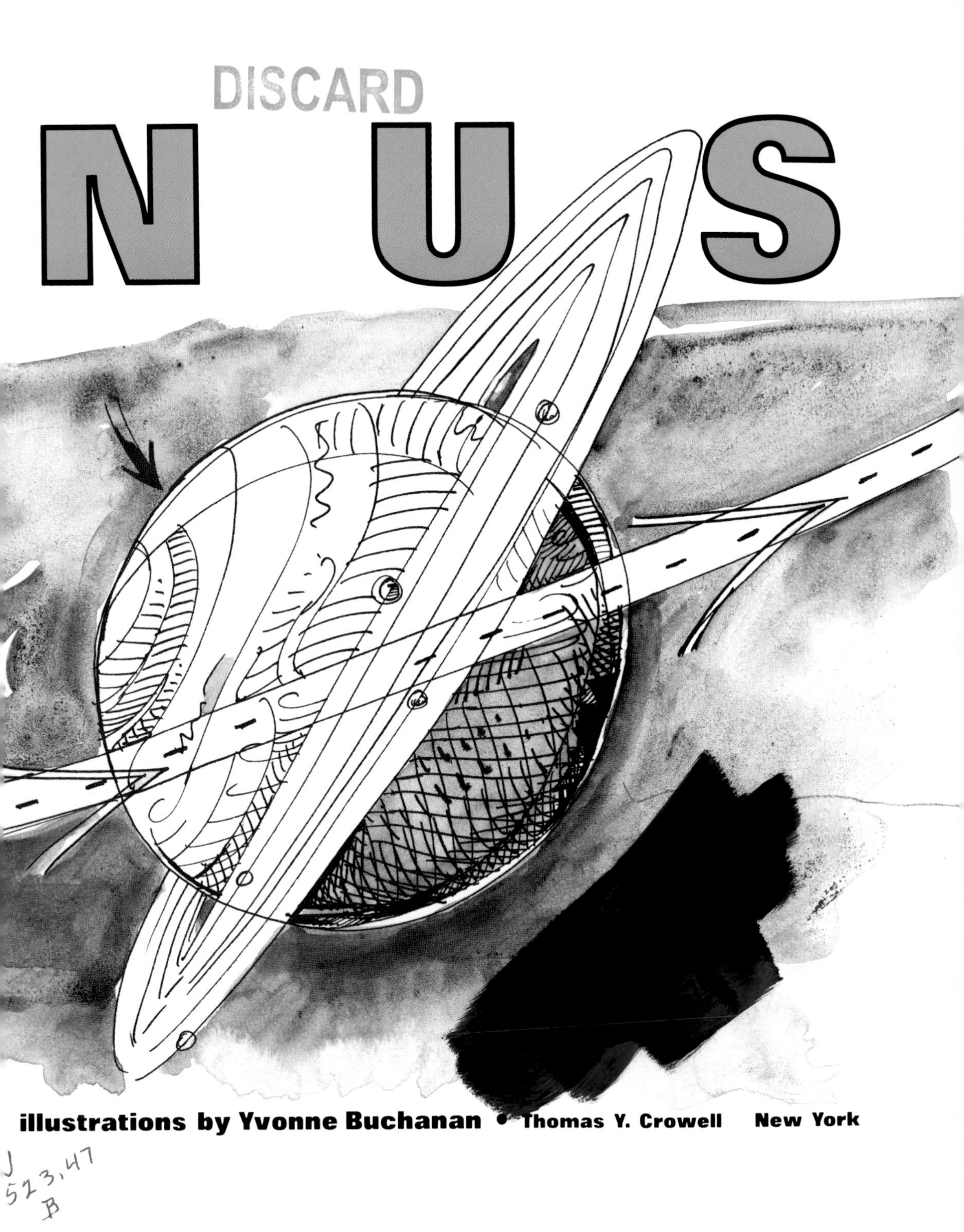

illustrations by Yvonne Buchanan • Thomas Y. Crowell New York

A special thank you to
Dr. J. Pieter de Vries, Manager, Voyager Flight Science Office,
for his perceptive judgment and valued assistance.

Uranus: The Seventh Planet
 Printed in the United States of America. For information address Thomas Y. Crowell Junior Books, 10 East 53rd Street, New York, N.Y. 10022. Published simultaneously in Canada by Fitzhenry & Whiteside Limited, Toronto.
10 9 8 7 6 5 4 3 2 1
First Edition
Photos courtesy of NASA/JPL

Library of Congress Cataloging-in-Publication Data
Branley, Franklyn Mansfield, 1915–
Uranus : the seventh planet / by Franklyn M. Branley; illustrated with photographs and with illustrations by Yvonne Buchanan.—1st ed.
p. cm.—(A Voyage into space book)
Bibliography: p.
Includes index.
Summary: Describes the physical characteristics, movements, satellites, and other features of Uranus, with an emphasis on recent discoveries from Project Voyager.
ISBN 0-690-04685-5 : $
ISBN 0-690-04687-1 (lib. bdg.) : $
1. Uranus (Planet)—Juvenile literature. 2. Project Voyager—Juvenile literature [1. Uranus (Planet)]
I. Buchanan, Yvonne, ill. II. Title. III. Series.
QB681.B73 1988 87-35046
523.4′—dc 19 CIP
AC

To Margaret with thanks
for the gift of 50 years

Other Voyage into Space Books

Saturn
The Spectacular Planet

Space Telescope

From Sputnik to Space Shuttles
Into the New Space Age

Star Guide

CONTENTS

(Color photo insert begins after page 24)

INTRODUCTION

If a pea were five football fields away from you, it would be very hard to see. That's the way Uranus appears to astronomers. It is the seventh planet from the Sun, about 1.8 billion miles away from it, and about 1.7 billion miles from planet Earth. Very little sunlight reaches Uranus, and very little of that light is reflected back to us. All the light we have received from Uranus during the past 200 years would add up to less than the light given off by a flashlight in one second.

So we can understand why a telescope was needed to discover the planet, and why it had to be a very good telescope.

This telescope, with a tube 7 feet long, was made by William Herschel in 1778. With it he discovered Uranus in 1781.

Uranus was discovered in 1781. That was 172 years after Galileo, the Italian astronomer, became the first person to systematically study the sky through a telescope. Galileo's telescope was small and not very powerful. After that, many more were made, and they were improved. In 1781 the very best telescope in existence was made by William Herschel, and he was using it when he discovered Uranus.

Herschel was born in Germany in 1738. His father was a member of the Hanoverian Foot Guards band. When Herschel was fourteen years old, he also joined the band. Later on he went to England, where he was to spend the rest of his life. Herschel was very interested in music, and he read all sorts of books about music; it was in those that he learned about the properties of sound waves, and this led him to read other books, some of which discussed light waves and the control of them by telescopes, lenses, and mirrors.

Herschel became more and more interested in telescopes, and soon he was spending a lot of time on astron-

William Herschel built many telescopes. This one, completed in 1783, had a tube 20 feet long. The tube could be raised and lowered and the entire instrument turned on rollers. Herschel discovered the moons Titania and Oberon using his 40-foot telescope.

William Herschel

omy and telescope making. Assisted by his brother, Alexander, and his sister, Caroline, who became a well-known astronomer in her own right, Herschel built several telescopes. The largest had a tube 40 feet long, but its size made it awkward to aim and focus. Herschel found that a smaller 20-foot telescope was more usable.

When he discovered Uranus on March 13, 1781, Herschel was using an even smaller, more compact telescope. It was 7 feet long. Uranus appeared as a very dim object among the stars of the constellation Gemini, the twins.

During the nights after March 13, Herschel watched the dim object change position against the background stars. It seemed to move as though it were a comet. And that's what Herschel believed the object to be. Later observations he and others made revealed that the movements were not like those of a comet—they were more like those of a planet.

George III, who was then king of England, encouraged Herschel and rewarded him. To show his gratitude, Herschel called the new planet George's Star. But the name was not accepted by astronomers. It was later changed to Uranus after the Greek god of the heavens.

The other planets are also named after Greek and Roman gods. Mars, Jupiter, Saturn and Uranus follow one another in ancestral order: Mars is the son of Jupiter, Jupiter is the son of Saturn, and Saturn is the son of Uranus.

The discovery of Uranus was startling. Up to that time the only planets people knew about were those whose movement among the stars could be seen easily without the help of a telescope. People believed that Saturn was at the edge of the solar system. All at once a planet twice as far away as Saturn was discovered—the size of the solar system was twice as great as people had thought.

Herschel could not see the planet very well, so he was able to learn very little about it. However, by watching the motion of the planet, he was able to calculate that it went around the Sun once in 84 years. Some six years later, Herschel discovered two of the satellites of Uranus.

Two more were found in 1851 by the English astronomer William Lassell. The fifth major satellite was discovered in 1948 by the Dutch American astronomer Gerard Kuiper.

Herschel's satellites are called Titania and Oberon. The other three are Umbriel, Ariel, and Miranda. Until

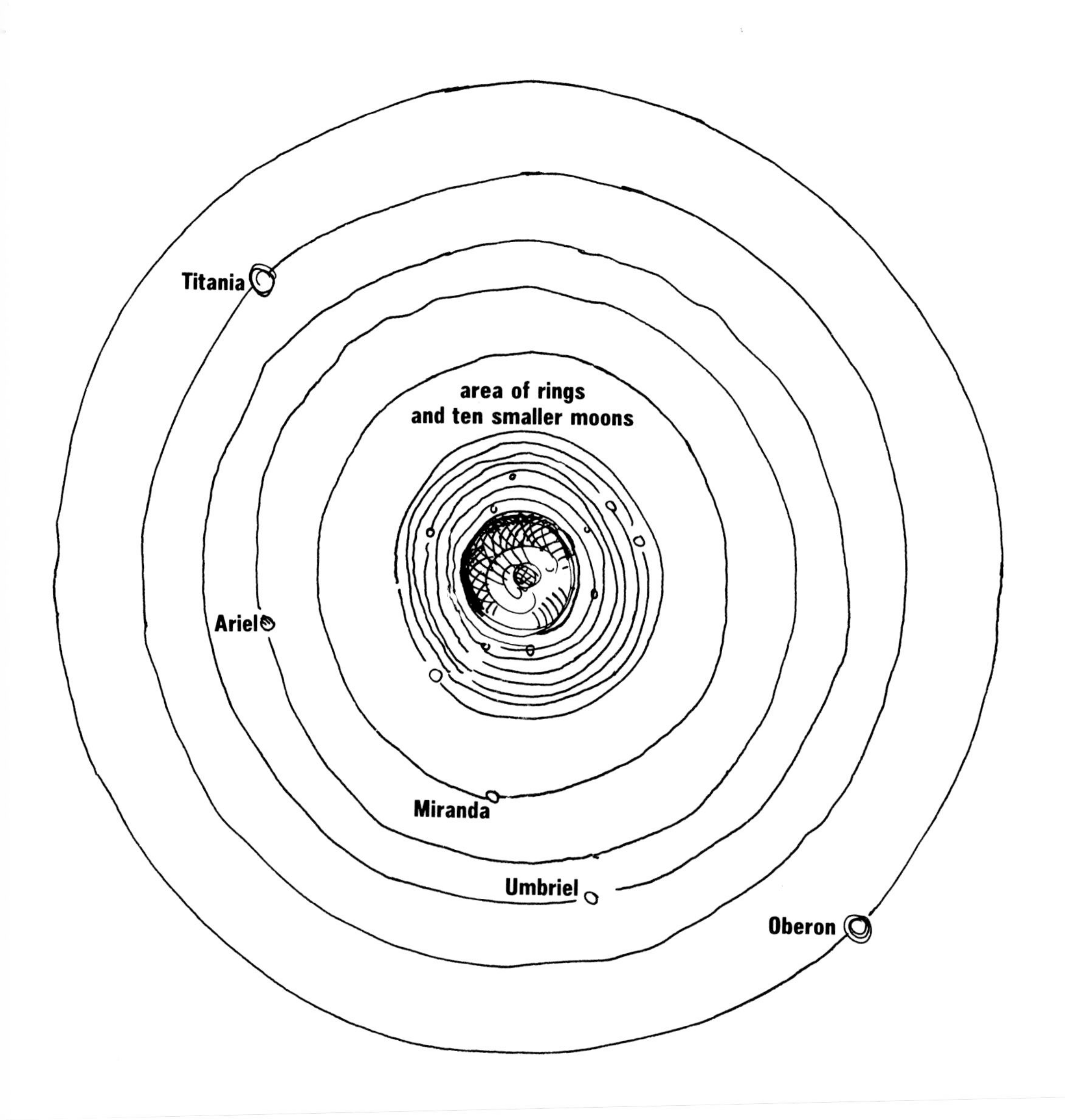
Titania
area of rings
and ten smaller moons
Ariel
Miranda
Umbriel
Oberon

1986 these five were believed to be the total number of satellites that revolved around Uranus. But in that year the planet probe Voyager 2 moved in close to the planet and found ten more satellites; so Uranus has fifteen satellites, and maybe even more.

Five large satellites belong to Uranus. In addition, ten smaller satellites were found between Miranda and Uranus. The planet has a system of at least eleven rings. The rings and satellites revolve around Uranus in essentially the same plane, nearly in line with its equator.

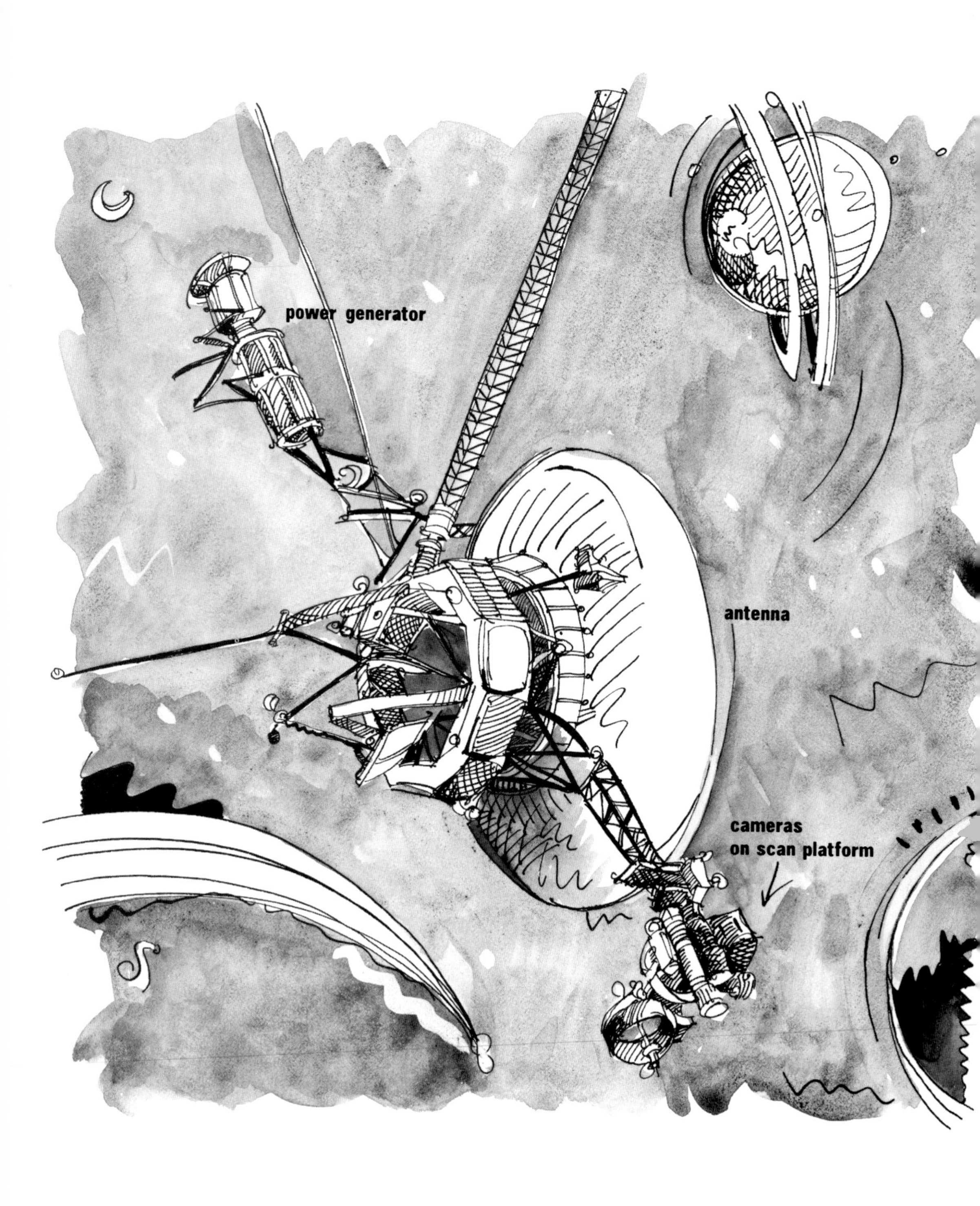
power generator
antenna
cameras
on scan platform

1. VOYAGER 2

Planet probes are unmanned spacecraft sent out to explore the solar system. They are loaded with cameras and instruments for gathering information, computers for data storage and control, radios for receiving commands, transmitters for sending data to Earth, jets to turn the probe as needed, and power generators to keep everything operating.

Voyager 2 was launched in August of 1977 and put on a course for Jupiter. The mission called for the probe to survey that planet two years later, which it did. The next target was to be Saturn, and in August 1981, four years after launch, Voyager sent us exciting information and

---

Voyager 2

pictures of Saturn and its rings. Its jobs were done. But Voyager was standing up very well; its instruments and transmitters were still operating. The close approach to Saturn and its strong gravitational field had speeded up the probe, and mission scientists decided to go for Uranus. A bit over eight years after its launch, Voyager moved in toward Uranus and gave us more information about the planet than had been gathered in the 200 years since Herschel had discovered it. For it to do this, the engineers had to solve many unusual problems.

Voyager 2 was launched in August 1977. Eight years later, in January 1986, it arrived at Uranus. After traveling over two billion miles, it was within 10 miles of the targeted position.

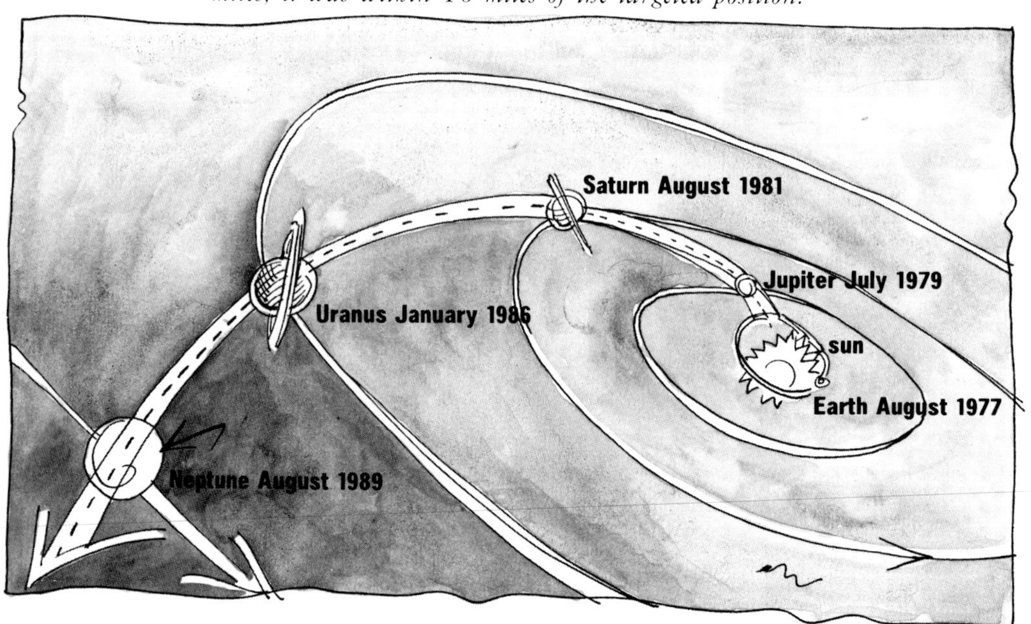

The problems became especially difficult because of the great distance: Uranus was 1.84 billion miles away at the time of Voyager's approach. That means it took 2 hours and 45 minutes for radio messages to travel from Earth to Voyager. Instructions for the probe had to be sent 2 hours and 45 minutes before they were to be carried out.

Another problem was that sunlight is so dim at Uranus, snapshot photos could not be taken. In order for Voyager's camera to collect enough light for anything to show up in the photographs, long exposures were needed. However, Voyager was traveling about 40,000 miles an hour, so it moved about 100 miles during a 10-second exposure. Unless something was done, the pictures would be blurred, just as they are if your camera moves as you're taking a picture.

The engineers programmed Voyager so the entire probe rotated slowly backward just enough to offset its motion forward. The effect was that the camera remained on target and was able to produce sharp time-exposure photographs.

The pictures were changed to radio signals that were then transmitted to Earth. The very weak signals were picked up by large dish antennas. They were amplified and changed back to visual images—several of which are in this book.

When Voyager 2 visited Uranus, it set a record, for Uranus was the most distant target surveyed by a planet probe. Now it is on a path that will take it toward Neptune.

Voyager's instruments may provide scientists with information until the year 2010, and maybe even longer. Eventually, though, the instruments will shut down—the probe will become inactive. But it will still be traveling through space. In the late 1990s it will reach the edge of the solar system—the region where the influence of the Sun just about disappears. Then Voyager will go on and on, through interstellar space, the empty region between the stars. Its path will take the probe toward Sirius; for millions of years it will be a space wanderer—alone in the universe.

2. URANUS: THE PLANET

Since its discovery, astronomers have not been able to study Uranus extensively. It is a dim object that has shown no surface features such as craters, channels, or other markings. It has seemed little more than a dull, bluish blur. Since we could see no markings, we could not be sure about how long it takes the planet to rotate.

Until 1975 the figure most often given for the rotation time of Uranus was 10.8 hours. After that date, a few astronomers identified clouds in the atmosphere. After tracking them, the astronomers variously believed the rotation period was between 15 and 24 hours. Changes in the magnetic field of Uranus, detected by Voyager, revealed that Uranus rotates in 17.24 hours (see Chapter 3, "The Magnetic Field").

All the planets except for Uranus rotate like tops—their axes are more or less straight up and down. All are tilted somewhat. For example, Earth's axis is tilted 23.5°, the tilt of Mars is 24°, Jupiter's is 3°, and so on.

The spin of Uranus is not at all like a top. It lies on its side—the tilt is 97°.

Why this should be so remains a mystery. However, many scientists believe that sometime during the early

---

The axis of Earth is tilted 23.5° from a line vertical to the ecliptic—the Earth's path around the Sun.

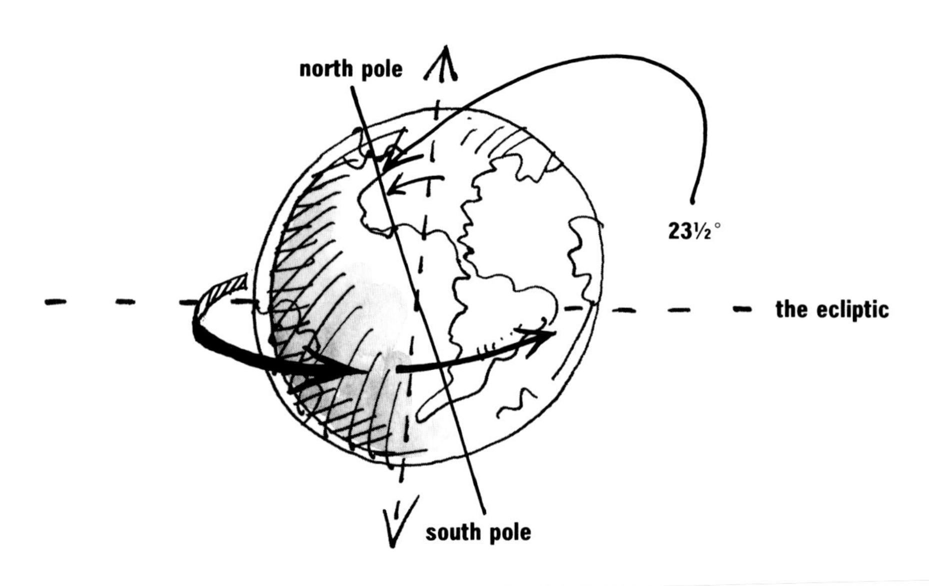

history of Uranus, a huge mass of material, perhaps of planet size, crashed into Uranus with such force that it pushed the planet over.

As a result, during its 84-year journey around the Sun, the poles are alternately toward the Sun and away from it. At the start, let's say, the north pole is almost in line with the Sun. Twenty-one years later, sunlight falls on the equator of Uranus, and after another 21 years, the south pole of the planet lines up with the Sun. For many years

The axis of Uranus is tilted 97°.

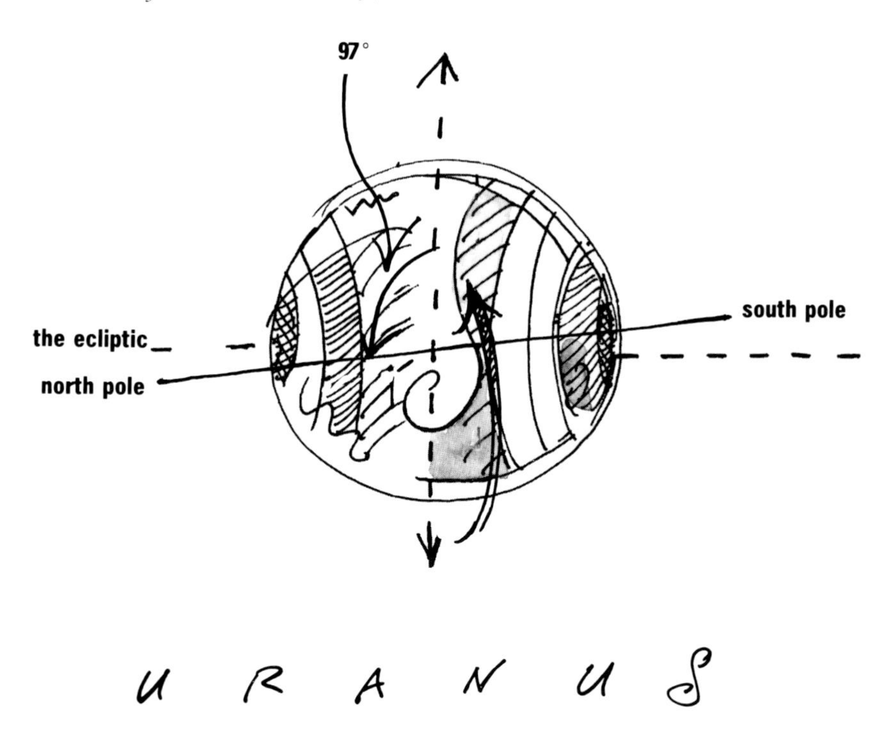

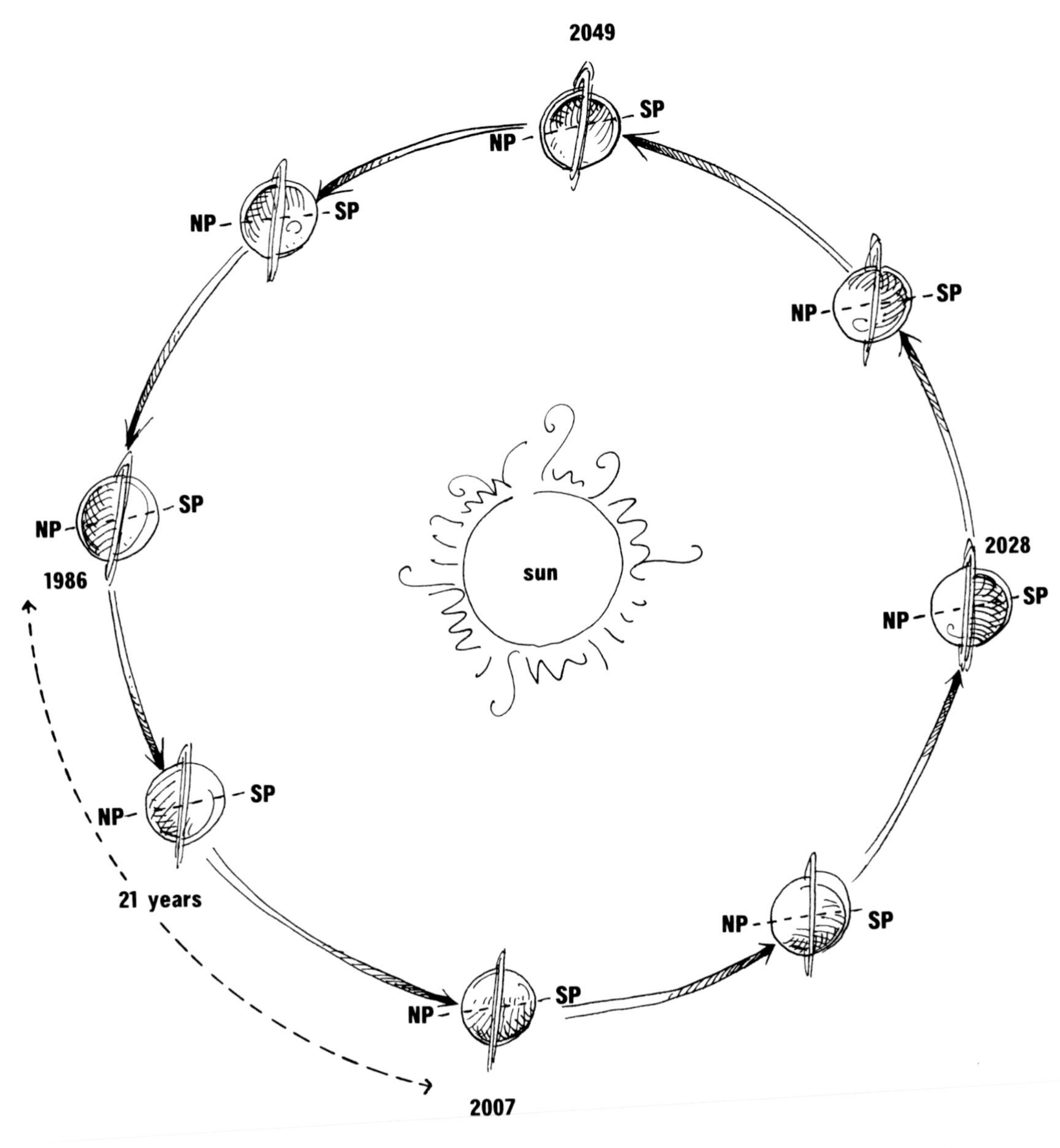

NP = north pole
SP = south pole

one polar region is in sunlight continually while the opposite one is in darkness.

The orbits of the satellites are almost in line with the equator of Uranus, so at times we see the planet as though it were the center of a bull's-eye, the rings and the orbits of the satellites making circles around the eye of the target. That's the way Uranus appeared when Voyager visited it, for the south pole was toward the Sun—and toward us.

In 1986 the south pole of Uranus was toward the Sun (and Earth). We saw the orbits of the satellites and rings as circles around a bull's-eye. In the year 2007 we'll see the orbits as vertical straight lines crossing the axis of Uranus. In 2028 the orbits will once again appear circular.

3. THE MAGNETIC FIELD

Earth is magnetic. You know this because a compass works on our planet. One end of the needle is attracted to the north magnetic pole. It is as though a bar magnet inside Earth extended through the planet, from the north magnetic pole to the south magnetic pole. The magnetic poles are about 11° removed from the geographic poles—the "ends" of the axis we revolve around.

It is believed that Earth is magnetic because beneath the surface there is a layer of semimolten material containing metals that conduct electricity. This semimolten material "flows" past solid layers. This movement sets up a flow of electricity, which generates the magnetic field that surrounds the planet.

Voyager discovered that Uranus also has a magnetic

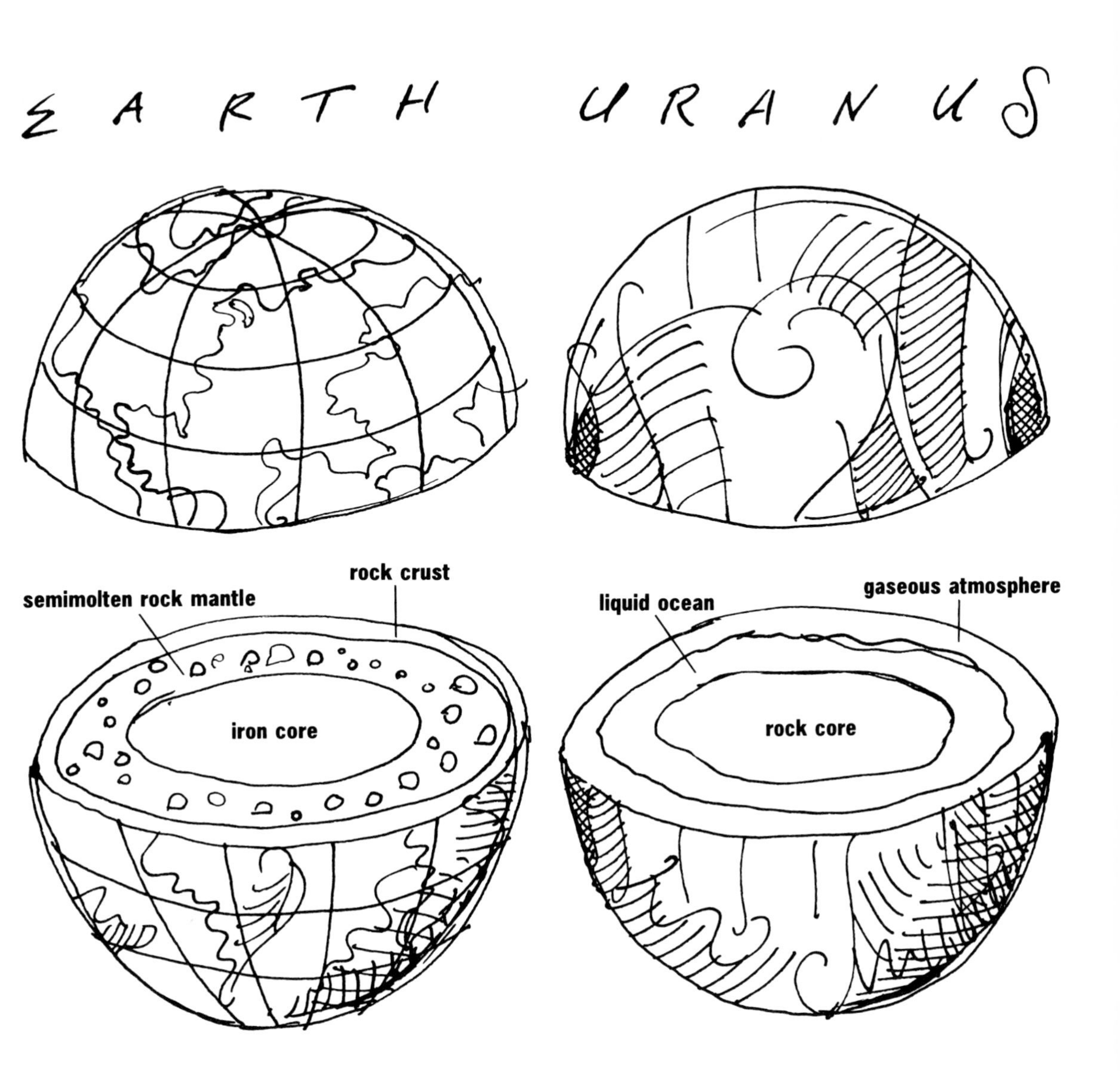

The structure of Earth is different from that of Uranus, probably because the two planets were not formed in the same way. Millions of comets may have contributed material to Uranus as the planet was forming. The material in Earth probably came mainly from the interior portion of the solar cloud, most of which became the Sun.

field. For any planet to have a magnetic field, scientists think it must have a semimolten or liquid layer of material. The liquid layer must be able to conduct electricity, and it must move or flow over a solid layer.

Like Jupiter, Saturn, and Neptune, Uranus is a "gas giant"—a planet made up primarily of gases. Uranus's atmosphere is 4000 or 5000 miles deep, and it contains hydrogen, methane, and some helium. At the center of the planet, some scientists believe, there is a core of solid material, largely dense rocks and metals. Between these two layers they think there may be a sea of "water" about 5000 miles deep. If that is so, the water would not be like

The magnetic pole of Earth is 11° from the geographic pole.

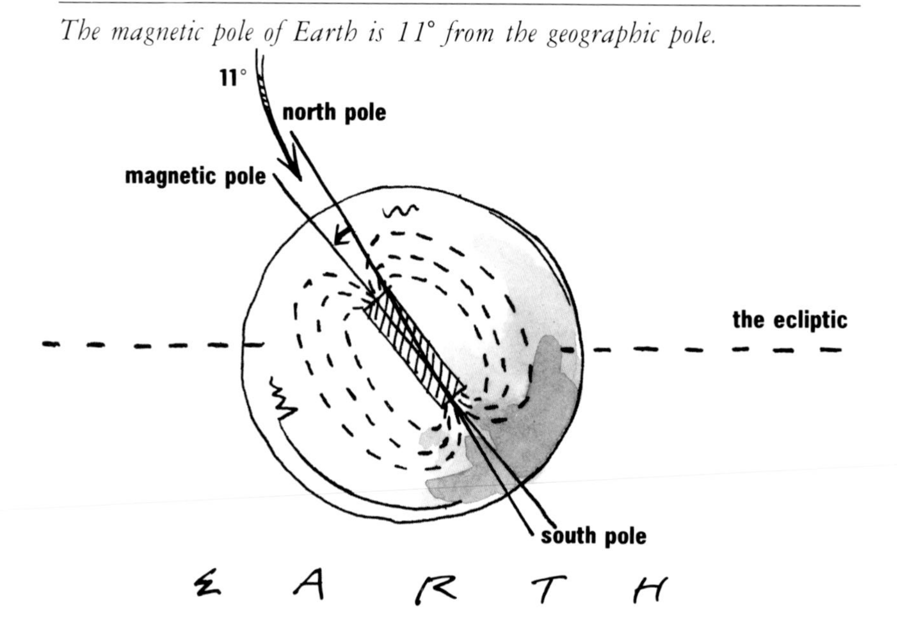

that in our oceans. The weight of the atmosphere exerts so much pressure, the molecules in the ocean would be broken apart into ions. (An ion is an atom or group of atoms that carries an electrical charge.) The ions are of the components of water, ammonia, and probably methane. These ions would make the ocean of Uranus electrical, and the churning action of the electrically charged liquid over the solid core of the planet may be what produces the magnetic field that has been detected.

Curiously, Uranus's magnetic field is tilted more than that of any other planet. The field is tilted some 60° from the geographic poles, or axis of rotation. So far there is no explanation of why this is so.

The magnetic pole of Uranus is 60° from its geographic pole.

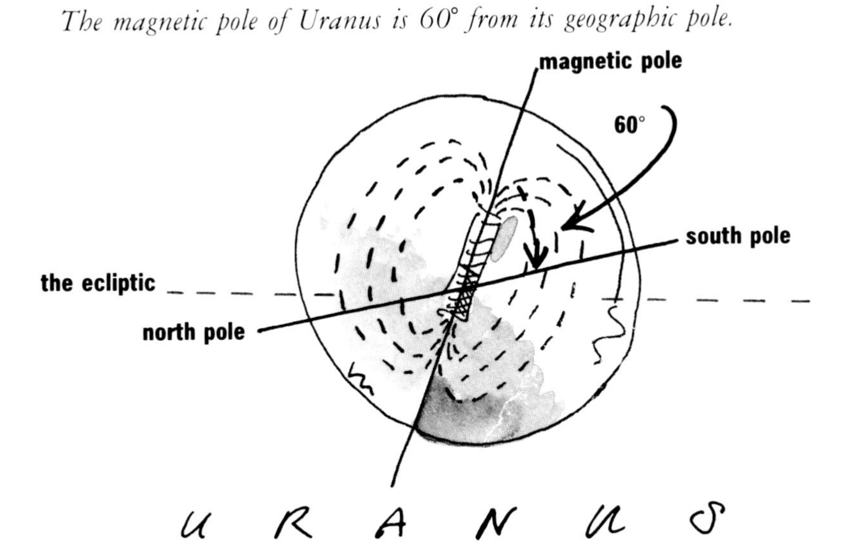

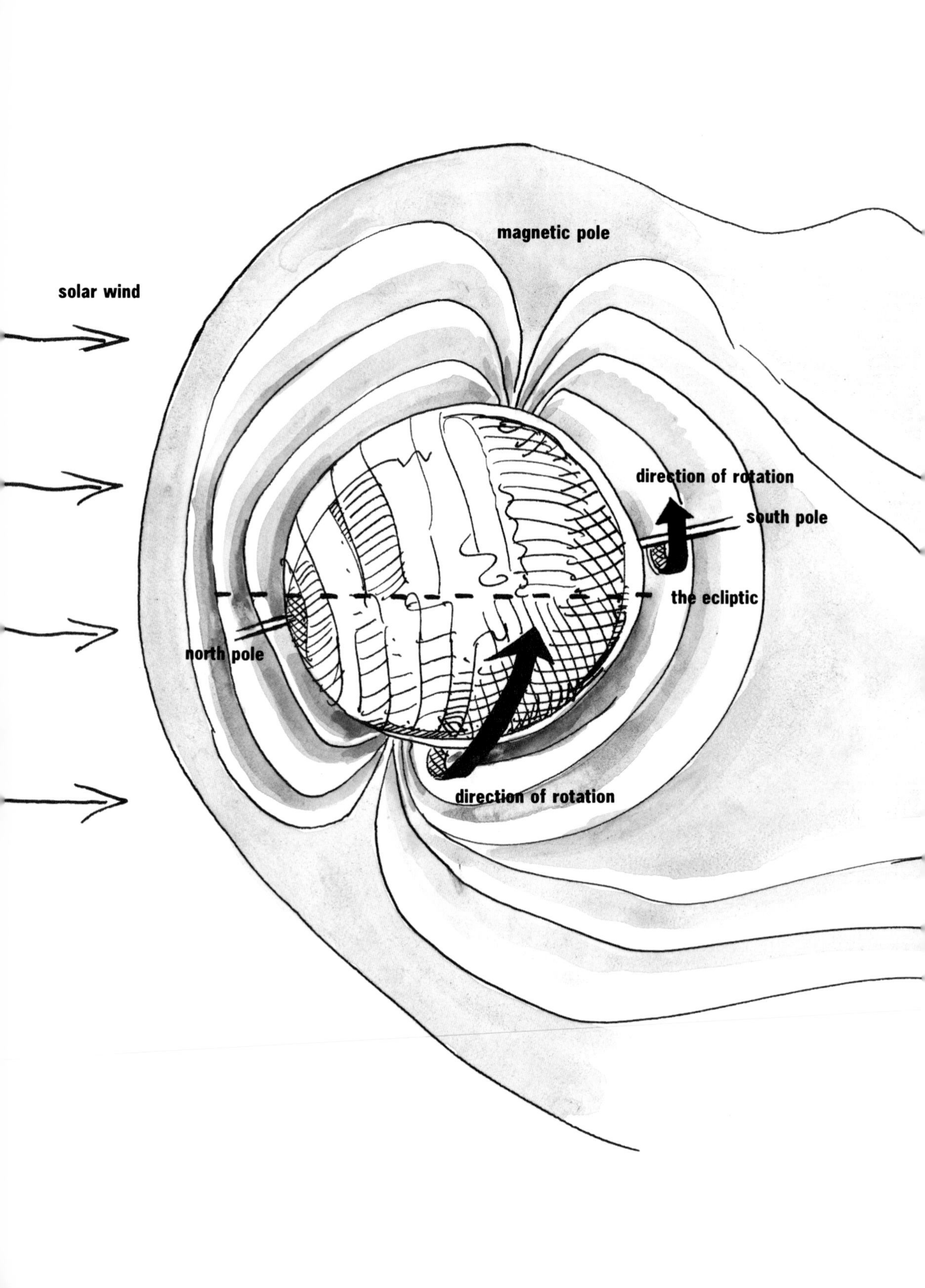
magnetic pole
solar wind
direction of rotation
south pole
the ecliptic
north pole
direction of rotation

Every planet with a magnetic field also has a magnetotail. Solar wind—particles that stream out continuously from the Sun—pushes the magnetic field away from the planet, and it streams outward in a "tail."

Uranus also has a magnetotail, but because of the far tilt of its magnetic field, the tail twists as Uranus spins. The twists match exactly the rotation of Uranus. Therefore, by measuring the changes in the magnetic tail of Uranus, scientists were able to determine the rotation period of the planet.

The solar wind, which is made of charged particles ejected from the Sun, pushes the magnetic field of Uranus into a long magnetotail. As Uranus rotates, the tail is twisted.

4. THE CLOUDS

Careful study of the small, dim telescopic image of Uranus enabled some observers to see bands in the atmosphere of the planet. Voyager 2 confirmed their existence. Uranus is an aquamarine planet, the soft blue color resulting from the presence of methane. All the light we see from Uranus, or from any other planet, is reflected sunlight. If all the colors of sunlight—red, orange, yellow, green, blue, and violet—were reflected equally, Uranus would appear white. We see it as a blue planet because the reds and yellows in sunlight are absorbed

---

At the top is Uranus as you would see it if you were aboard Voyager 2. It is a soft-aquamarine-colored planet; the color is produced when methane gas absorbs the reds of sunlight and reflects the blues and greens. At the bottom is Uranus as a computer sees it using false colors to enhance details. The reddish area above center is the south pole. The bands around it are probably produced by movements in the atmosphere. The bright streak at the lower left edge is not a real image. It was produced in the process of enhancing details.

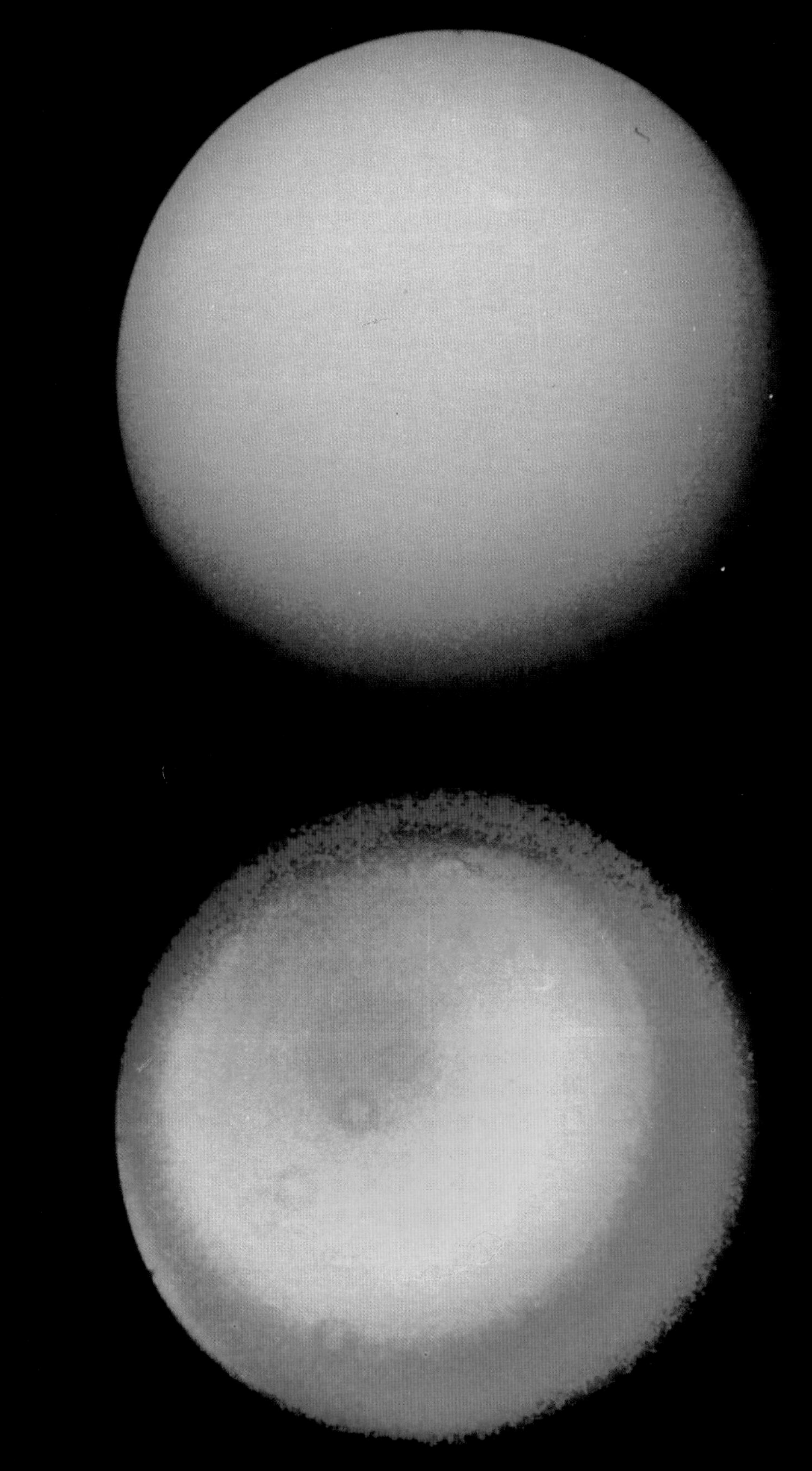

Uranus as it would be seen from a spacecraft about to land on Miranda. The planet is 65,000 miles away. The deep canyons of Miranda are seen in detail (see also pictures on pages 32 and 33), and the dark rings of Uranus are clear.

Close-up of Ariel (see also photo on page 38). You can see the filled-in valleys at lower right. The bright streaks are probably crater walls reflecting sunlight.

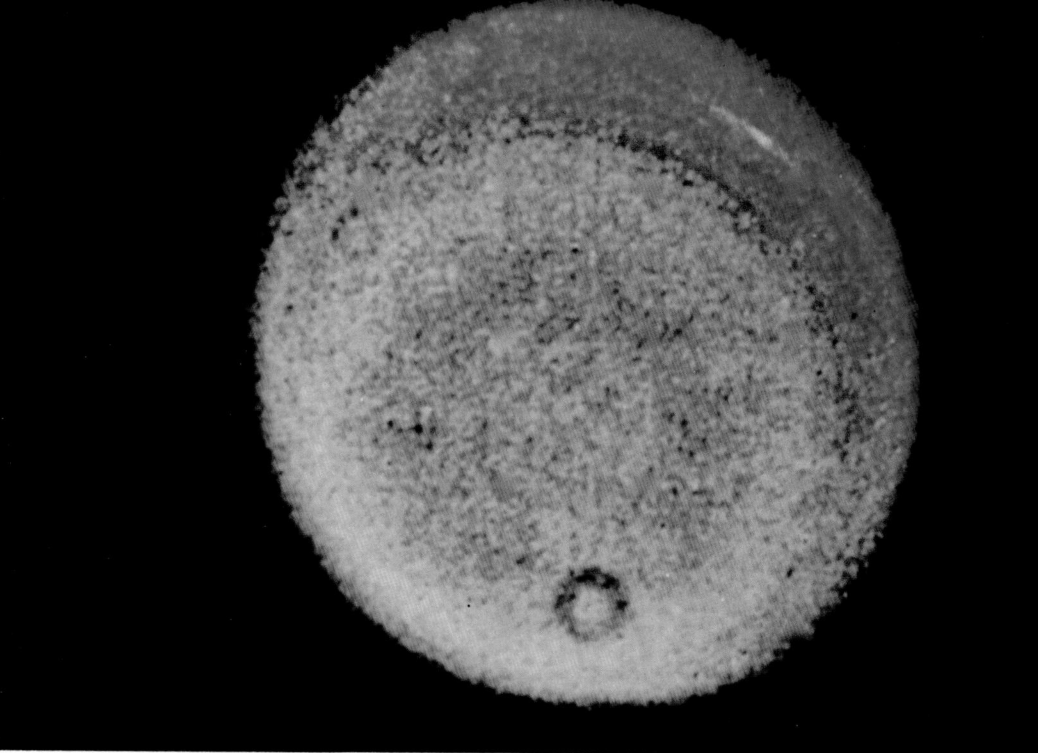

Uranus as only a computer can see it. False color exaggerates the details. The light-pink patch is a cloud, one of three that were identified. The color bands may result from layering in the atmosphere, or from molecules that vary in structure in different parts of the atmosphere. This is one case where the computer poses questions rather than delivering answers.

As Voyager left Uranus and headed toward Neptune, it looked back and took this picture of the planet some 600,000 miles away. The soft blue-green at the edge fades into white, caused by haze in the upper atmosphere.

Nine of the eleven rings of Uranus can be identified. The top ring is the widest. Beneath that is a group of three, then a pair of rings. At the bottom is another group of three. The faintly colored bands were produced in the computer while the photo was assembled and enhanced.

Antennas 210 feet across located in California, Spain, and Australia picked up the weak signals sent out by Voyager 2 over a distance of nearly two billion miles.

(removed) by methane gas; only the greens and blues are reflected—together they make the planet aquamarine.

Voyager revealed very few clouds. If more exist, they must be deep below the outer atmosphere. However, the probe did detect the bands or stripes that extend around the planet. The bands are believed to be produced by convection currents. The lighter regions appear where relatively warm material flows upward in the atmosphere, the darker areas where colder material flows downward.

The few clouds that were seen indicate there are strong winds blowing on Uranus. On Earth, upper winds may blow some 100 to 200 miles an hour. On Uranus the winds appear to blow at two or three times that speed.

The deep atmosphere of Uranus is unbelievably cold. It is so far from the Sun, very little energy is received. The temperature, measured at a level in the atmosphere where atmospheric pressure is equal to that on Earth's surface, is around 300° F below zero. One would expect that the south pole, which is presently toward the Sun, should be much warmer than the rest of the planet. But this is not the case. The small amount of solar heat is spread from one place to another, perhaps by the strong winds that appear to blow constantly.

5. THE OCEAN

Underneath Uranus's atmosphere some scientists think there may be an ocean, thousands of miles deep. The tremendous pressure from Uranus's atmosphere would be great enough to solidify the liquid of its ocean. However, the water is extremely hot—hot enough to remain liquid in spite of the extreme pressure. In fact Uranus's ocean is probably several thousand degrees Fahrenheit.

Without the intense pressure from Uranus's atmosphere, the superheated ocean would boil and be converted into steam. Increased pressure on a liquid raises its boiling temperature. Here on Earth, water at sea level boils at a higher temperature than it does at Denver, Colorado, which is a mile above sea level. That is because at sea level pressure is higher. On Uranus pressure is so high, the liquid temperature can be several thousand degrees and still not boil.

Some of the water in the ocean may have come from the interior of the planet by way of extensive volcanic eruptions. But it is believed the larger part came from showers of millions of comets that may have collided with the planet and melted. That would explain the large amount of water. It would also explain why there are large amounts of somewhat heavier elements, such as carbon, oxygen, and nitrogen, in the Uranian makeup. Also, it would explain the relatively high density of Uranus as compared to Jupiter and Saturn.

The inner planets—Mercury, Venus, Earth, and Mars—all have high densities (density being the amount of material packed into a given volume of space). Based upon water, which has a density of 1, the densities of Mercury, Venus, and Earth are all greater than 5; and the density of Mars is 3.9. The densities then drop off steeply—Jupiter's density is 1.33, while the density of Saturn is 0.69. As you can see, density drops rather steadily as distance from the Sun increases. Then, there is an abrupt increase in the densities of Uranus (1.18), and Neptune (1.56), which are beyond Saturn.

The reason may be found in a theory of planet formation. The four inner planets, plus Jupiter and Saturn, very likely formed from a great solar cloud. The largest part of that cloud became the Sun. Small amounts of leftover dense material became the inner planets, and less dense matter nearer the outer edge of the cloud became Jupiter and Saturn.

Matter from that same cloud probably made up part of Uranus and Neptune as well. However, much of the matter in the far-out planets may have come from a comet cloud that surrounded the original sun-making cloud. That comet cloud is believed to still be the source of comets.

We know that comets contain large amounts of ice and rocky material, as well as carbon and nitrogen. Millions and perhaps billions of comets may have bombarded the core of Uranus, adding material to it. The collisions and the force of gravity packing the accumulated material together would have generated the extremely high surface temperature, much of which is still retained, unable to escape through the dense atmosphere. The ice would have melted to form the superocean of Uranus.

Fast-moving comets that did not crash into one of the planets may have had their paths deflected enough by the planet's gravity to return them to the comet cloud. These comets, plus great masses of dust not yet consolidated into comets, make up the vast cloud of billions of comets as extensive as the solar system itself.

6. THE SATELLITES

The five major satellites of Uranus extend from 80,700 miles—the location of Miranda, the closest satellite—to 362,500 miles, the location of Oberon, the most distant. You may remember that our satellite, the Moon, is 239,000 miles from Earth.

All five satellites are small, especially for satellites of such a large planet. The smallest is Miranda, 293 miles across, and the largest is Titania with a diameter of 982 miles. The diameter of our Moon is 2,160 miles.

In 1986 Voyager 2 arrived at Uranus just one minute ahead of the time scheduled five years earlier. It discovered ten more satellites in the space between Uranus and Miranda. All are very small—they average about 45 miles across.

Voyager approached Uranus from the direction of the Sun (and Earth), and so it headed toward the south pole of the planet. Because the orbits of the satellites of Uranus extend outward from its equator, making the planet appear as the center of a bull's-eye, Voyager was not able to pass from one satellite to another, crossing the orbits. It passed closest to Miranda, and we have excellent photos of that satellite. But pictures taken of the other satellites revealed interesting information as well.

In 1986 Voyager 2 approached Uranus at its south pole. It flew through the band of satellites and rings, and so was able to photograph them during only a brief time.

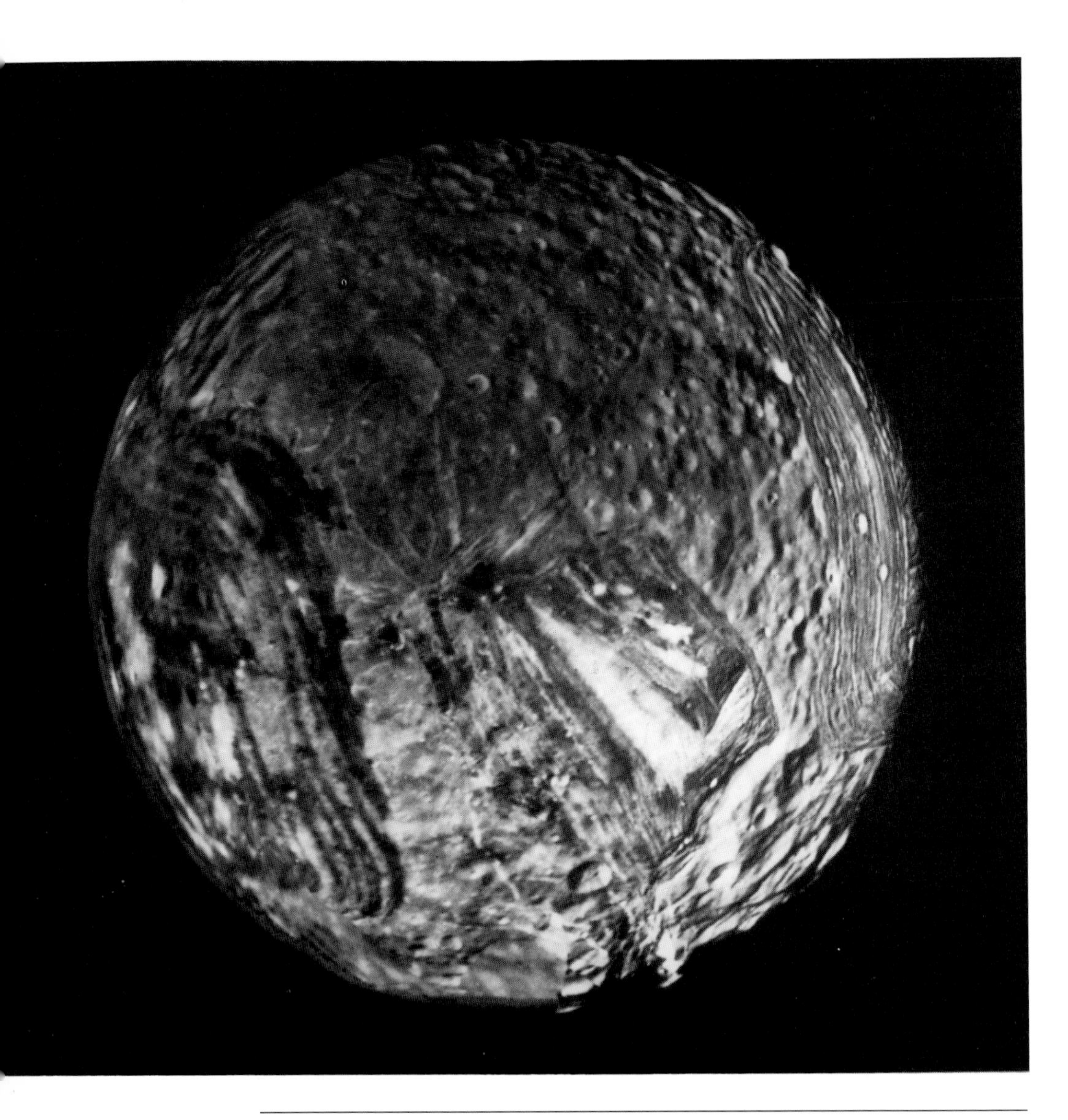

Startling Miranda, about 293 miles across, is the innermost of the five large satellites. Several photos were assembled to make this picture. Notice the variety of features—smooth regions containing craters, areas with deep canyons where the surface has parted, and sections pushed together making long highlands perhaps 3 miles high. Whatever happened on Miranda, it must have been shattering.

MIRANDA is made largely of ice and rock mixed with frozen methane. Craters were expected. But most unexpected were other features, such as deep valleys that remind one of Mars, and highlands broken by deep craters and high cliffs.

The V just below center in the first photo and in the lower left in the close-up photo is made of long ridges of highlands and lowlands. In the lower left and upper right of the first photo are similar streaks. Between these patches are sections that look much like the craters of the Moon.

A close-up of Miranda taken from 26,000 miles, and including an area 140 miles across. The white V at the lower left shows canyons running at right angles to one another. The rugged highlands and canyons at the upper right also make a right-angle turn. The craters were probably made by impact after the streaked regions had formed.

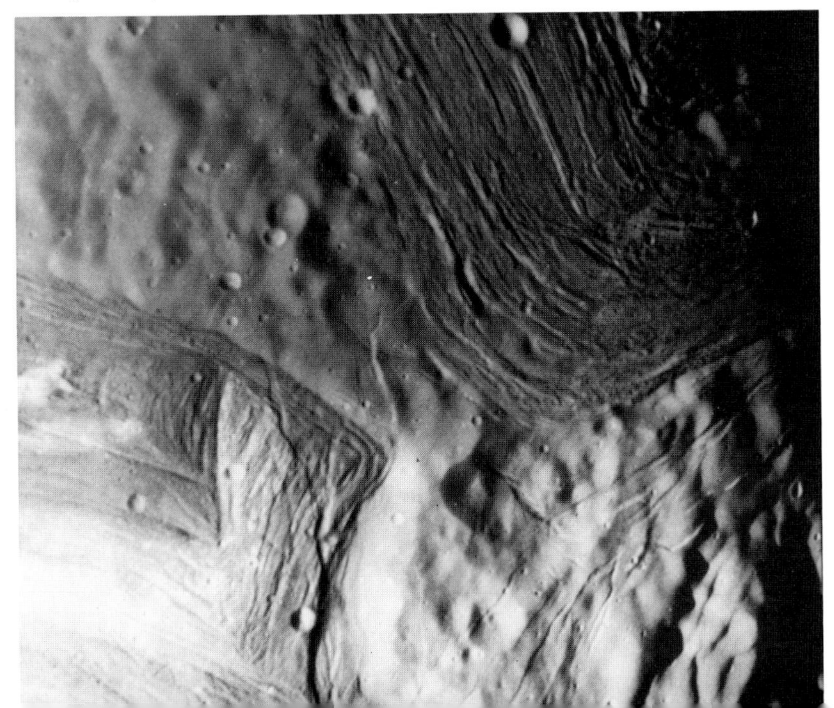

Miranda may have started out as an iceball, much like many other planetary satellites. After rocky material had settled inward, the satellite may have been bombarded by a huge mass—perhaps as large as itself. The impact was great enough to break up the satellite into large and small chunks. Gradually these loose chunks came together again. Some chunks were placed so the ice side was outward. These are the lighter-colored crater-marked areas that we see now. Other chunks were turned so the darker, rocky interior sections were turned outward. These are the dark, streaked sections that are so unusual. This may have been the history of Miranda, or the explanation for its strange appearance may be entirely different. Scientists have observed, however, that huge crevices, such as those found on Miranda, are more common on smaller celestial bodies. This is because there is less gravity to hold them together.

Each of the other four larger satellites has distinctive features. All appear to be made largely of ice, which is a good reflector, yet all reflect very little light. They are thousands of times dimmer than the dimmest stars we can see with our eyes alone. It is believed that methane may be mixed in with the ice. Methane (CH_4) contains carbon

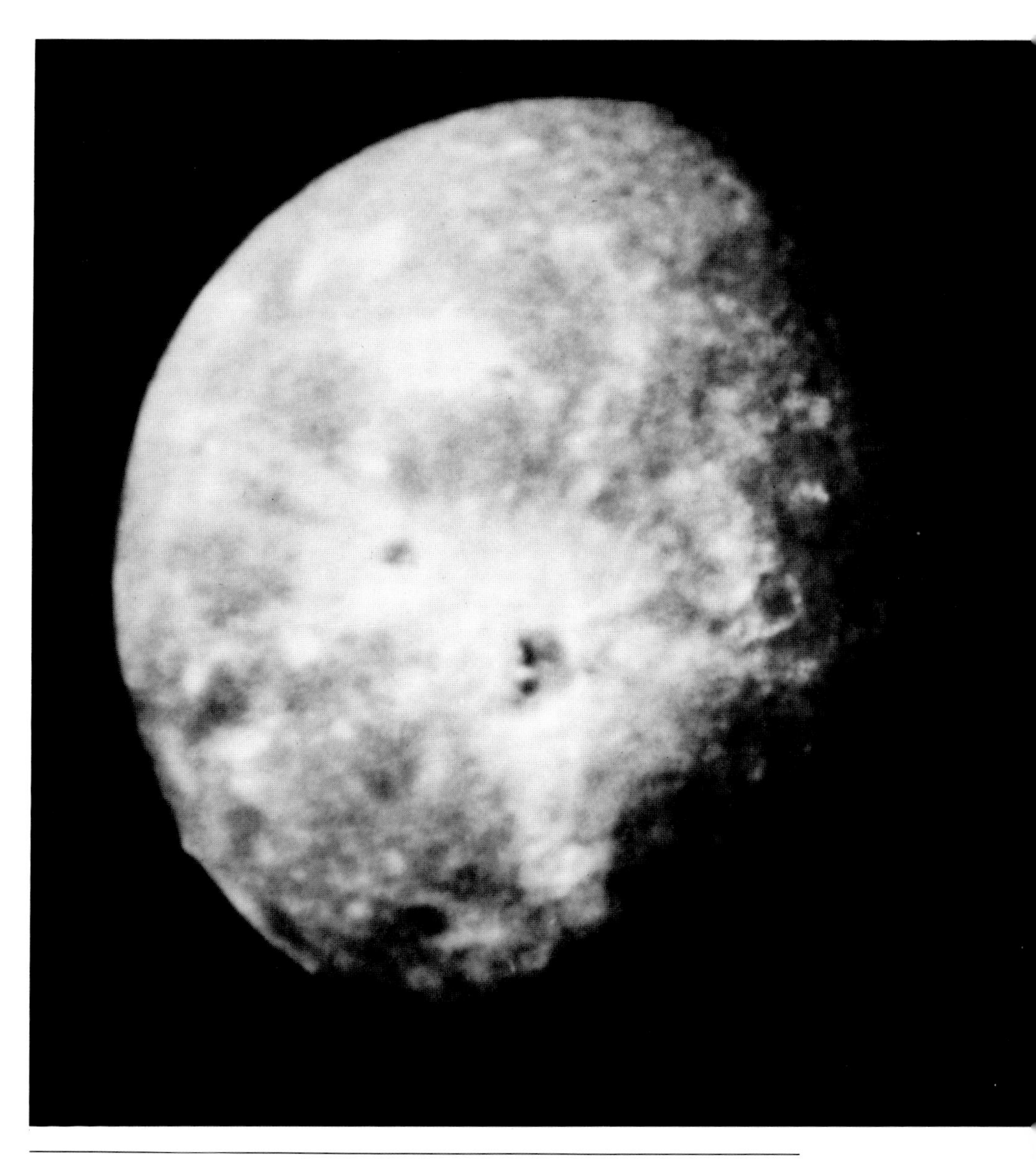

Oberon, the outermost satellite, is 362,500 miles from the planet. It appears to be an ice satellite covered with craters and material sprayed from the craters to the surface. Dark patches in some craters may have been produced when lava was ejected from beneath the surface.

and hydrogen. The methane may have broken down, releasing hydrogen and leaving black carbon—a very poor reflector.

OBERON is covered with craters, which means there was heavy bombardment during its early history. Many of the craters appear to be filled in, which means lava of some kind must have welled up from beneath the surface. The lava may have been rocky material and also methane-produced carbon molecules. The slight bulge at the lower left of the satellite is a mountain that is about 4 miles high.

TITANIA, which is a bit closer to Uranus, is also crater covered. But as you can see, there are long streaks on the surface. These are deep valleys hundreds of miles long and 45 miles wide. They may have been produced as the satellite cooled and contracted, or glancing collisions with huge masses may have gouged them out.

ARIEL and UMBRIEL are nearly the same size, and their orbits are quite close together. Yet, curiously, the two satellites appear to be quite different. Ariel shows long, deep valleys that may have formed when the crust

Titania, the largest of the satellites, is about 982 miles across. The surface has been heavily bombarded, as the craters show. The bright streak near the center is caused by sunlight and perhaps frost. At the bottom is a crater 125 miles in diameter, with a valley, produced by a later event, cutting across it. The crater at the top is 180 miles across.

Ariel is pitted with craters, many about 6 miles across, and streaked with valleys and highlands. The surface appears to have cracked open, perhaps as the crust settled after first expanding. Valleys at the lower right appear to have been filled in. Perhaps material from the interior flowed to the surface. The diameter of Ariel is about 720 miles.

Umbriel, about 728 miles across, is covered with craters, making it textured like the skin of an orange. The curious white patch at the pole may be the walls of a large crater reflecting sunlight.

expanded, contracted, and cracked. Here and there the valleys seem to be filled in. Lava ejected by volcanoes may have flowed into them. Ariel has certainly been an active satellite.

On the other hand, Umbriel has been inactive. There are no deep valleys or signs of volcanic activity. Why some satellites show signs of volcanic activity and others do not remains an unsolved mystery.

Umbriel's entire surface is covered with craters, many of them 50 to 100 miles across. As on the satellites of other planets, the craters were very likely caused by bombardments during early stages of formation. A peculiar feature is the white area atop the satellite. What it might be is still a puzzle.

None of the ten smaller satellites inside Miranda's orbit were seen well enough to discover features on their surfaces. These smaller satellites move within the ring system of Uranus.

7. THE RINGS

In 1977, scientists knew, Uranus was to pass in front of a star. By watching carefully, astronomers would be able to see slight changes in the starlight as it shone through the atmosphere of Uranus. Those changes would enable astronomers to learn something about the depth and structure of the atmosphere.

The best place to watch this event was from the area of the Indian Ocean. And, to be sure clouds did not interfere, the ideal location was high in an airplane above the ocean.

Astronomers from Cornell University carried their telescopes and other instruments aboard a high-flying plane. To get ready for Uranus's passage in front of the star, the astronomers turned on their equipment early.

They were amazed to discover that minutes before Uranus's atmosphere had reached the star, the starlight dimmed and then brightened—not once, but five times. Perhaps small satellites were the cause. But the same dimming occurred again as Uranus moved away from the star. The only formation that could produce such changes was a ring system. There must be five rings around Uranus.

A year later, other observations were made, and they revealed there were at least nine rings in the system. In 1986 Voyager discovered there were two more. So Uranus has a system of at least eleven separate rings, and there may be others. They lie inside Miranda's orbit, in the same area where the ten small satellites are found.

The rings escaped discovery earlier because they are very dim. They are made of material that reflects very little light. Also, the sunlight received at Uranus is only 1/400th as much as the sunlight that Earth receives.

For the most part the rings are made of very small particles, although in some cases the particles may be as big as boulders or even a house. Between many of the rings there appear to be clouds of dust particles that have

Bands of microscopic dust—perhaps as many as 100 bands—are dispersed among Uranus's rings.

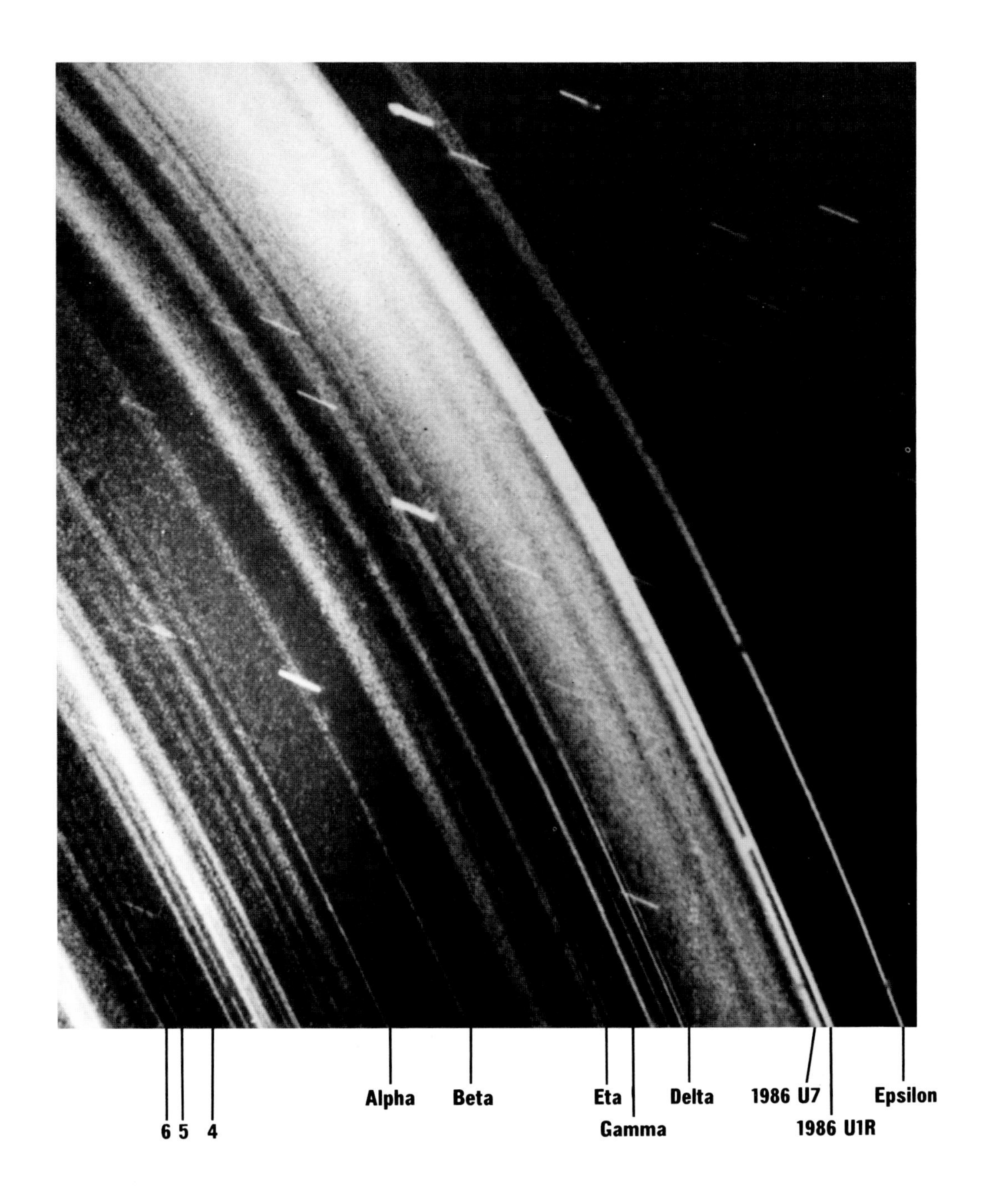
6
5
4
Alpha
Beta
Eta
Gamma
Delta
1986 U7
1986 U1R
Epsilon

leaked away from the rings, or have not joined to make rings. All the rings are made of separate particles so, in a way, each ring is made of millions of tiny satellites—all holding a formation. Eventually, many of the separate rings may converge to make a single wide ring. Or the particles may be pulled to Uranus by gravity, causing the rings to disappear.

During the brief period that Voyager passed by Miranda and above the atmosphere of Uranus, millions of bits of information were sent to Earth. Scientists will continue to study the information to learn all they can about Uranus, its rings, and its satellites. Already the planet has turned out to be much more interesting than anyone suspected. Too bad William Herschel wasn't able to see the results of Voyager's trip through the Uranus system. He would have been amazed by the discoveries.

APPENDICES

Uranus and Earth

	URANUS	EARTH
Rotation	17.24 hours	23.93 hours
Revolution	84.018 years	365.26 days
Orbital velocity	4.23 miles per second	18.51 miles per second
Tilt of axis	97°	23.5°
Tilt of magnetic field	60°	11°
Mass (Earth = 1)	14.5	1
Diameter	31,763 miles	7927 miles
Volume (Earth = 1)	63	1
Density (Water = 1)	1.18	5.52
Distance from Sun (in millions of miles)		
Mean	1786	93
Greatest	1871	94.5
Least	1702	91.5
Satellites	15 (or more)	1
Atmosphere (main parts)	Hydrogen, Methane	Nitrogen, Oxygen

Satellites of Uranus

NAME	DIAMETER (IN MILES)	DISTANCE FROM URANUS (IN MILES)
Oberon	947	362,500
Titania	982	271,100
Umbriel	728	165,300
Ariel	720	118,600
Miranda	293	80,700
1985 U1	106	53,440
1986 U5	37	46,790
1986 U4	37	43,430
1986 U1	50	41,070
1986 U2	50	40,020
1986 U6	37	38,970
1986 U3	37	38,410
1986 U9	31	36,790
1986 U8	31	33,430
1986 U7	25	30,940

FURTHER READING

Beatty, J. Kelly. "A Place Called Uranus," in *Sky and Telescope*, April 1986, pp. 333–37.

Branley, Franklyn M. *Saturn: The Spectacular Planet.* New York: Thomas Y. Crowell, 1983.

———. *Jupiter: King of the Gods, Giant of the Planets.* New York: Lodestar Books, 1981.

———. *The Nine Planets*, rev. ed. New York: Thomas Y. Crowell, 1978.

Chaikin, Andrew. "Voyager Among the Ice Worlds," in *Sky and Telescope*, April 1986, pp. 338–43.

Gold, Michael. "Voyager to the Seventh Planet," in *Science 86*, May 1986, pp. 32–39.

INDEX